At the Symposium in 1986, Alicia Rios gave one of the most entertaining talks I have ever attended on any subject. Her wit and enthusiasm were infectious. She made me realize that olive oil was not just something to cook with, but a precious gift. With a history as old as time, it has a variety of organoleptic experiences that no other fruit, apart from the grape, can ever match. When I tasted the different oils she presented, it was like my first taste of really fine wine. I was converted.

So I started, just for my personal interest, to research the subject. What began as curiosity has ended in a book! I discovered at the outset that all the pieces were there, but no one had pulled them together. It is truly amazing that with so much written about the grape, there was so little about its historical peer — the olive. The subject has proved vaster than I ever imagined when I embarked on my quest.

What I have come to realize is how important the seemingly humble olive was in the formation of empires. How deeply rooted in legend and folklore it is and how many communities rely on its very existence for their livelihood.

The joy for me is that many of my favorite places are linked by the olive and it has been delightful to journey to countries, not just as a tourist, but with a specific mission — a mission which has taken me to venues I would never otherwise have visited and introduced me to people I would never have met. The more I have become immersed in my theme, the more compulsive it has become.

The first seeds of my interest in things gastronomic had been sown by my godfather, who taught me the joys of good food, good wine and classical music. For this I owe him eternal thanks. To have revealed at such a tender age the finer things of life, is an education indeed.

Since those first tentative steps on the culinary road, I have proceeded along a fascinating path of discovery. I started traveling seriously after leaving university, at which point I began to appreciate the delights of simple, well prepared food — a broiled fish in a Greek island taverna, a boeuf en daube in a French truck drivers' cafe, tapas and a glass of fino in a village bar in Spain, the street food in Egypt and Morocco, fresh pasta in Italy, or a breakfast of fresh bread and creamy yogurt in Turkey.

You get near to the heart of a country when you sample its food, and you truly begin to understand the people when you wander into their markets and see how they shop. You may visit the monuments and marvel at the architecture in the Mediterranean to put you in touch with its glorious past, but the present is lived round these market stalls and the harbors where the latest catch has been landed.

Writing this book has given me the opportunity to deepen this education. It has been a wonderful, exciting odyssey. It is a strange path that leads one to discover a topic of interest and to find that it stimulates you sufficiently to want to share the excitement with other people. If I can persuade you in the following pages to follow my voyage of discovery, I promise you will not be disappointed.

THE OLIVE AND THE ANCIENT WORLD

"The whole Mediterranean, the sculpture, the palms, the gold beads, the bearded heroes, the wine, the ideas, the ships, the moonlight, the winged gorgons, the bronze men, the philosophers, all of it seems to rise in the sour, pungent taste of these black olives between the teeth. A taste older than meat, older than wine. A taste as old as cold water."

LAWRENCE DURRELL, *PROSPERO'S CELL*

Detail from an amphora in the British Museum

T he history of the olive weaves its way through myths and legends, wars and treaties, commerce and culture, theology, medicine and gastronomy. Its precise origins are lost back in the mists of time. However, the olive could be viewed as a genuine benchmark of civilization itself. While primitive people were hunting and scavenging, while they were nomadic, they didn't plough to cultivate crops; only when they were settled and creative, did they have time for planting trees and building palaces and temples. So it's not surprising to find the olive figuring as prominently as it does in Egyptian, Greek and Roman history.

Who first discovered the uses of the olive, and how to extract its unctuous oil no one really knows. The remarkable fact is that the green olive is inedible in its natural state; it needs to be treated with water or brine to remove the bitter glucosides. So did people come to discover the delights of the olive by eating ripe, black fruit or did they one day gather olives which had fallen into the sea at the water's edge?

There are indications that it was first cultivated in Syria by a Semitic race, 6,000 years ago, for records show that Ancient Palestine was famous for its olive groves, and exported oil to the Ancient Egyptians.

The Room of the Olive Press at Knossos

The Bible, not surprisingly, contains many references to the culinary and religious uses of olives and olive oil. In the *Book of Genesis* the dove sent out from the ark by Noah returned with an olive branch. Here it became the great symbol of peace, indicating the end of God's anger. And its recognition by Noah suggests that it was already a well known tree.

Ancient Roman vase with olive motif

The greatest religious significance of olive oil is documented in the *Book of Exodus*, where Moses is told by the Lord how to make an anointing oil of spices and olive oil. During consecration, holy anointing oil was poured over the heads of kings and priests, and it is still used today in the Roman Catholic Church, at baptisms and in the last rites for the dying.

The sacred use of olive oil also extended into the preparation of food used during sacrifices, such as pure wholewheat flour kneaded with olive oil, which is mentioned in the *Book of Leviticus*.

Another reference showing how important olive oil was in Biblical times is given in the *Book of Judges*. The trees wanted to elect a king to rule over them and so they chose the olive tree, but the tree refused, saying "and give up my oil by which gods and men are honored and go to sway over the trees?"

The olive has had great significance for every religion, though, and the Qu'ran contains this wonderful passage, "Allah is the light of the heavens and of the earth. His light is like a niche wherein a lamp is to be found, the lamp is in a glass cover which seems to be a twinkling star; it is lit thanks to a sacred tree: the olive tree, which has an oil so clear that it would give light even if no spark were put to it."

There is speculation that the olive may have appeared first west of the Nile Delta, as the Ancient Egyptians certainly knew how to grow olives. Or it may have been introduced into Egypt by the Hyksos, a Semitic tribe, who invaded Upper Egypt around 1650 BC, and remained for about 100 years before being driven out by the Egyptian king, Amosis. Its commerce is mentioned in Egyptian records, olives are depicted in tomb paintings, and branches have been found in sarcophagi. Garlands of olive branches even crowned the head of Tutankhamun.

It was the goddess Isis, wife of the supreme god Osiris, who was praised for giving the Egyptians the necessary knowledge to cultivate the trees and produce oil. But their olive oil production never subsequently matched that of the Greeks and Romans. Though the Egyptians used olive oil for cooking and medicine, it is noted in Greek records as being of poor quality, and so they imported olive oil from Syria and Crete.

Inscribed earthenware tablets dating back to 2500 BC, from Crete at the time of King Minos, are some of the oldest surviving references to olive oil. They mention different types of oil as well as the uses made of it. The palace at Knossos was the center of economic life and contained a chamber known as The Room of the Olive Press. Hundreds of the great amphorae, some standing nearly five feet high, in which olive oil was stored, can still be seen today.

At the height of Cretan trading with the islands of the eastern Mediterranean and Egypt, these amphorae, some of which contained wine as well as olive oil, represented Minoan currency.

Following the collapse of Cretan power in 1200 BC, the Phoenicians became the masters of trade in the Mediterranean, Tyre being the commercial capital. About this time, the Phoenicians taught the Greeks to use oil as a source of light and to make the terracotta lamps necessary to burn oil. By the 5th century BC, the Greeks and Phoenicians were major exporters of olive oil and the olive had been introduced to the shores of the western Mediterranean countries, on the course of their trading routes to Spain.

The olive must have occurred naturally in Greece, however, because the Greeks gave it their own name — *elaia*. If it had been introduced from Syria, they would have adopted the Hebrew, *zayit*, or *zeit*, as it became in the Arabic countries of North Africa.

Greek legend records a famous story which accounts for one

Olea Europaea species is divided into three sub-species: Olea Europaea Euromediterranea: sativa; oleaster; Laperrini: typica var; cyrenaica var; maireana var; Cuspidata: varied.

The Olea Euromediterranea sativa or Olea sativa HOFFM and LINK is also known as the cultivated olive tree; there are literally hundreds of varieties and, like the grape, the variety grown depends on the climate, the soil and also whether the olives are to be pressed for oil or preserved for eating. With table olives, it is firmness and fleshiness of fruit which is important, whereas olives grown for pressing must have a high oil content. There is no difference in the varieties of green and black olives, for all olives are green at first and turn black when fully ripe, changing through a beautiful spectrum of colors; rose, wine-rose, brown, deep violet, deep chestnut, reddish black, and finally violet black. Some olives, such as the Spanish Manzanilla and the French Picholine, are tastier when picked green, others, such as the famous, tiny Niçois and the Greek Kalamata, are best when fully ripe.

The olive tree cannot tolerate extreme cold or damp but can survive lengthy periods of drought. So it is chiefly to be found between the latitudes 25° and 45° North, and it especially flourishes in the Mediterranean climate, with its mild winters and long hot summers.

The tree can grow to a great height, as much as 50ft., but most are pruned, so they remain at about half this, to facilitate picking.

Its trunk, which is smooth and gray when young, becomes twisted and knarled with the years. It takes about five to eight years before an olive tree will bear its first fruit and then it can go on producing for years and years. Olive trees have a great tenacity of life because when the main trunk dies, new shoots sprout up around its base, eventually growing into a new tree. It is not unusual, for instance, for olive trees to live to 600 years or more. So it is not impossible that the trees at present in the Garden of Gethsemane could be those under which Christ prayed, the night before he died.

New trees can be grown from seed but they take years to reach the fruit-bearing stage, the most common method of propagation being to take cuttings from mature trees. Some varieties or cultivars have developed resistance to cold, others to pests and drought. The advantage of propagation from cuttings is that the new tree will carry these resistances.

The tree is an evergreen and its leaves live for about three years, before dying and making way for new ones. Olive leaves are paired opposite each other down the branches. They are single and undivided, rather like a willow leaf, lance-shaped, shiny and leathery. The upper surface is dark green and the lower surface appears to be a silvery-green, because it is covered with minute scales. This is why in the wind, olive trees seem to shimmer in a silvery haze.

The tree blooms in late spring with clusters of white flowers. Depending on the variety, there can be anything from ten to over forty flowers in a cluster, but only one in every twenty flowers will become an olive. Even though olive trees are self-pollinating, it is very difficult for the flowers to achieve fertilization at the best of times, but the weather is the prime enemy. Rain at blossom time can be disastrous. As a result, fruit setting is erratic but it is improved by planting another variety of olive tree for cross-pollination.

Between June and October fruition takes place. This is the time during which the pit (endocarp) hardens and the pulp (mesocarp) fills out. The flesh is encased by the skin (epicarp) and as the olive ripens the epicarp changes from green through violet and red to black. Six to eight months after the blossoms appear, the olives have maximum oil content; they are black and fully ripe.

The olive is a pit fruit, like the cherry, and it is the slender pit encased inside the fruit which contains the seed of the tree itself. Botanically this type of fleshy fruit encasing a stone is known as a drupe, from the Greek, Δρυππα, meaning an overripe, wrinkled olive.

Olive trees are mostly found in arid terrain; they need little rain and can survive in the poorest soils because the roots bore deep into the earth in search of what little moisture they require. In these conditions, they will produce a mediocre harvest. However, with care and attention the tree is far more productive and bears fruit every year, rather than every other year.

Growing olives is now big business; olive oil is a world

commodity, so modern methods of cultivation are being used increasingly, much like other vital crops — and these days on most commercial estates trees are fertilized, pruned and irrigated.

In early autumn and spring, the soil in the groves is ploughed and weeded and the trees are pruned. Pruning is a most important but labor-intensive process; by thinning out the growth on the crown of the tree, the fruit-bearing branches can be exposed to sun and air. The removal of vigorous but sterile branches and suckers, which appear in abundance at the base of the tree, means that the nutrients in the soil will be better utilized. A tree can yield between 35 and 45 pounds of olives depending on how it is tended. Think of the work involved in going over every tree, laboriously and carefully one at a time by hand. There can be no short cuts, and there is no possible means of mechanized pruning.

Like all fruit trees, the olive is subject to attack from fungi and insects. Its particular enemy is the olive moth, whose caterpillars will happily eat their way through leaves and buds. These enemies need to be controlled, and so spraying with pesticides may also be one of the duties of the olive producer. However, some producers choose to grow their olives organically because of concern over contamination of the fruit by pesticides.

The olive grower's year is certainly a full and active one. Apart from pruning and fertilizing, they have to harvest and press the olives.

The desired object of all this work is, of course, to produce a good crop of olives, but the bounty which the olive tree provides is not in its fruit alone; there is almost no part of this eternal tree which cannot be utilized.

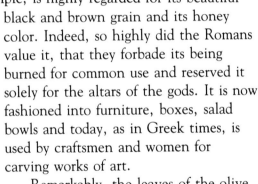

Olive wood, for example, is highly regarded for its beautiful black and brown grain and its honey color. Indeed, so highly did the Romans value it, that they forbade its being burned for common use and reserved it solely for the altars of the gods. It is now fashioned into furniture, boxes, salad bowls and today, as in Greek times, is used by craftsmen and women for carving works of art.

Remarkably, the leaves of the olive tree have always been regarded as curative when boiled in water and taken as an infusion. This infusion is recommended for sufferers of hypertension and those with heart conditions, as well as being an effective diuretic. Even the residue of pits and skins after pressing, is used in many mills as fuel, and may also be used as fertilizer for the trees. Can there be another tree with such a history, with such folklore surrounding it and with such a multiplicity of uses?

THE FRUIT OF THE TREE

"He causeth the grass to grow for the cattle, and plants for the service of man: That he may bring forth food out of the earth; and wine that maketh glad the heart of man, and oil to make his face shine . . ."

PSALMS.

Going out to buy olive oil today, you are presented with an exciting but perhaps bewildering choice in supermarkets, wine shops and specialty stores. There are virgin oils, and pure oils from Spain, Italy, France, and Greece. There are single estate oils and national brands as well as own brands from the large multiples. There are emerald green oils and golden yellow oils. Bottles and cans of all shapes, sizes and designs. So where does someone start to make a selection?

Olive oil, like wine, has an enormous diversity of flavors. The taste, color and aroma are dependent on the country of origin, the soil on which the trees are grown, the variety of olive and the method of harvesting. There are also different qualities which are reflected in the price, but the information on the label will help you make an informed choice.

Cold pressed, extra virgin olive oil is the best you can buy. Virgin oil will vary in taste and color from year to year because it is the pure, unadulterated juice of the olive. Unlike wine, however, olive oil does not improve with age. It is at its best the year after pressing. The color gives no indication of its quality because if the olives are green when they are harvested the oil will be different from the oil of black, ripe olives. Olive oils have a range of colors, from deep green to light gold, and as many flavors, from pungent and peppery to the light and fruity. It's all a matter of personal taste, but you may find that the extra virgin oils are best used for dressing salads and adding uncooked to certain dishes such as soup because of their powerful flavors. They come into their own when used in regional dishes where the flavor of the local olive oil gives the dish its intrinsic taste; a dish of ratatouille from Provence, a bowl of pasta from Italy, a luscious tagine from Morocco — they all owe their distinctiveness to the local olive.

Pure olive oils, which are the refined oils blended with virgin oil to give flavor, color and aroma, are lighter in flavor and are excellent for making dishes where you don't want the taste to be so strong as to dominate the other ingredients — as with mayonnaise, or for frying.

National brands of pure olive oil such as Carbonell, Sasso, Berio, Bertolli and Cypressa or supermarket own brands, because they are refined and blended will have the same flavor each time you buy a bottle.

To understand the differences in virgin oils and refined oils, it helps to know what happens to the olive once it has been plucked from the tree.

For those who know and love the Mediterranean, summer is the time associated with the riches of fresh figs, aromatic herbs and

Young tree

Separating olives from the twigs and leaves

Now, as then, once the olives are crushed, the resulting paste is spread onto loosely woven hemp mats. These are stacked, up to fifty at a time, interspersed with metal discs, which help to distribute the pressure evenly in the press, where they undergo hundreds of tons of weight in an hydraulic press. Today where this form of pressing is still in use (but is becoming rarer), the mats are usually made of nylon rather than esparto. The mats allow the oil to drain through while holding back more solid matter. At this stage, the oil contains some of the fruit's own water. Traditionally, the liquid then flowed into tanks or troughs where the oil was

Hand picking at harvest time

Early North African oil mill

31

allowed to settle. Being less dense than the water, the oil rises to the surface and can be decanted off. Centuries ago the oil was subsequently stored in sealed amphorae, which protected it from bright sunlight.

Sadly, the romance and simplicity of old-style milling and pressing has now all but gone, and when you visit a modern mill you see very little but bright stainless steel equipment worked by electricity and controlled by computer, occupying only a few square yards of space. The washed olives disappear into a hopper at one end and the oil emerges at the other, the whole process being continuous. However, one can't deny that progress has brought about a superior quality olive oil, produced under the most hygienic conditions. It really is a totally natural product untouched by human hands.

Today, there is no need to spread pounds of paste on mats and stack them in a press. Once the olives are crushed the paste moves into a cylindrical trough, where blades turn the mixture over and over to form a homogeneous mass. This is called malaxation.

The oil is separated from the paste by means of centrifugation, which simply means spinning the paste round at high speed. These modern methods are all adaptations of the old principles, simply taken to the degree of sophistication necessary to ensure the profitable production of olive oil in today's world.

Olive oil produced by these methods is known as first cold pressed olive oil. It is made usually from green, unripe olives which have been pressed once and to which no heat or chemicals have been applied and it takes about 10 pounds of olives to make 1 quart of oil. Except for centrifuging and perhaps filtering, the oil remains untreated. No other vegetable oil is edible just being pressed. All other seed oils have to be treated first because they contain toxins or are not suitable for human consumption in their natural state.

First cold pressed olive oil is a totally pure product and it retains all its natural flavors. Because of this it is possible to recognize the origin of the oil, just as wine can be identified by its intrinsic organoleptic qualities.

These first cold pressed oils are virgin oils, graded according to their acidity in standards set down by the International Olive Oil Council. Approximately 65–90% of oil extracted is virgin olive oil and is ready for consumption.

Olive oil produced by the same natural methods but which because of excessive acidity, color or flavor is not fit for consumption, is known as lampante virgin olive oil. Such oil needs to be refined, which is done by neutralization, bleaching and deodorization.

In the production of first cold pressed olive oil, solid matter known as residue remains and still contains a certain amount of oil. This is mixed with water and heated to 160°F, and is pressed again, yielding an oil of lesser quality and higher acidity. A somewhat similar grade is obtained from bruised fruit. Pressing can be repeated once or twice, yielding each time a poorer grade. At the last press, oil is extracted from the residue with carbon disulphide, more recently with trichlorethylene to yield olive residue oil.

These crude residue oils are very highly colored, have a strong taste and are very high in acidity. They cannot be marketed simply as olive oil. In this state they are unfit for consumption but, like lampante virgin olive oil, can be treated for sale as "refined olive-residue oil" — or, when mixed with virgin olive oil, as "refined olive-residue oil and olive oil." Olive residue oils that are too high in acidity are used for industrial purposes such as wool combing in the textile industry, in the manufacture of toilet preparations and in making high quality castile soap.

The process of refining the olive oil can be a chemical or a physical process. In the chemical process, caustic soda is added to the oil to draw out the fatty acids. In the physical process, which is being used increasingly, olive oil is put in a drum, which is heated up in a vacuum, to about 360°F. This evaporates the fatty acids and they are drawn off. In this way the oil is neutralized and deodorized.

Another physical method of refining is to use a centrifuge. This is a cheaper process and therefore oils with a very high acidity are centrifuged first. This drives off the greater part of the free fatty acids and the oil can then be treated in the vacuum, to remove the rest.

THE PROCESS OF PRODUCING
OLIVE OIL

It takes 5 kg of olives to make
1 litre of olive oil

Washing

Trituration

The
olive
mill

Olive paste is
spread on mats

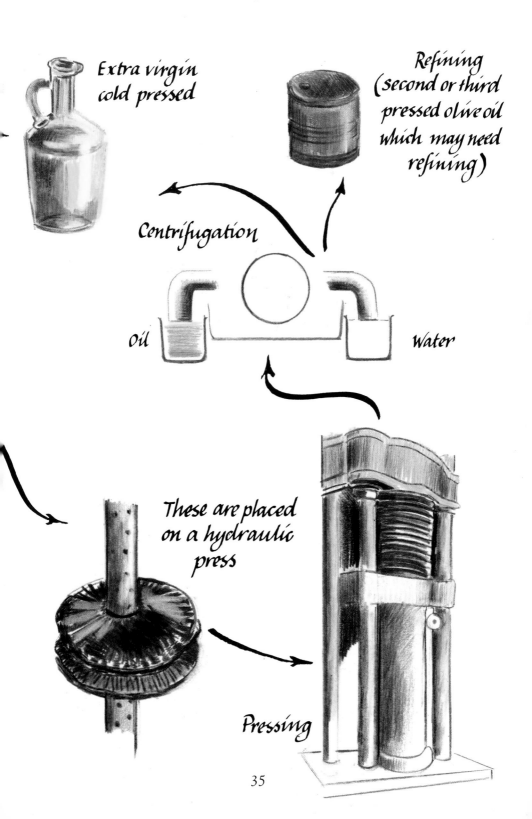

Extra virgin
cold pressed

Refining
(second or third
pressed olive oil
which may need
refining)

Centrifugation

Oil

Water

These are placed
on a hydraulic
press

Pressing

Bleaching to remove the color is the next process in refining. The oil is first run through a sand, whose properties attract that part of the oil which gives it its color. Then it is filtered to remove any particles of sand which may remain.

In Spain, Italy, France, Greece and Tunisia, research is going on constantly to improve all the various processes of milling and refining. In 1970 a Center for the Improvement and Demonstration of Olive Oil Extraction Methods was set up in Cordoba, Spain, for the benefit of all olive oil producers. There is also an experimental oil mill at Seville in Spain called the Instituto de la Grasa y sus Derivados. Places such as these strive to make olive oil production more efficient and cost effective in the face of rising labor costs. The development of the continuous centrifuge method of extraction, for example, has obviated the need for pressing mats which are used in the hydraulic presses and which, because they are made today from synthetic materials derived from petroleum, have become expensive to produce.

Olive oil is a world commodity like wheat, sugar or cocoa and as such its production and labeling is controlled by international legislation and it is this which provides the consumer with assurances on quality.

The first International Olive Oil Agreement was drawn up in 1959, and is administered by the International Olive Oil Council (IOOC), whose headquarters are in Madrid. This council sits twice a year and comprises all olive oil producing countries and olive oil importing members. The EEC is a member, along with Algeria, Egypt, Libya, Morocco, Tunisia and Turkey.

The current Olive Oil Agreement came into force in 1979 as an extended and amended version of the 1959 Agreement, and it basically seeks to promote the expansion of the world market by encouraging the improvement of olive oil productivity. It also sets down standards for all of the categories of olive oil. A label reproducing the IOOC emblem, the use of which is governed by regulations, guarantees to the consumer the purity and quality.

The designation "olive oil" is given to the oil obtained solely from the fruit of the olive tree, to the exclusion of oils obtained using solvents or re-esterification processes, or any mixtures with oils of other kinds.

GRADES OF OLIVE OIL AS DESIGNATED BY THE INTERNATIONAL OLIVE OIL COUNCIL

VIRGIN OLIVE OIL is the oil obtained from the fruit of the olive tree solely by mechanical or other physical means under conditions, and particularly thermal conditions, that do not lead to alterations in the oil. Further, the olive has not undergone any treatment other than washing, centrifugation and filtration.

Virgin Olive Oil fit for consumption as it is, is classified into:

EXTRA VIRGIN OLIVE OIL. This is virgin olive oil of absolutely perfect taste and odor with an acidity of less than 1%.

FINE VIRGIN OLIVE OIL. Virgin olive oil of absolutely perfect taste and odor having a maximum acidity of less than 1.5%.

SEMI-FINE OLIVE OIL (or ordinary virgin olive oil) is virgin olive oil of good taste and odor, having a maximum acidity of 3%.

Virgin olive oil not fit for consumption as it is, is designated **VIRGIN OLIVE OIL LAMPANTE** (lamp oil). This is an off-taste and/or off-smelling virgin olive oil. It is intended for refining or for technical purposes.

REFINED OLIVE OIL is olive oil obtained from virgin olive oils by refining methods.

OLIVE OIL or **PURE OLIVE OIL** is oil consisting of a blend of refined olive oil and virgin olive oil.

Finally, there is olive residue oil, which is a crude oil obtained by treating olive residues with solvents, and intended for subsequent refining for human consumption. It is classified as **REFINED OLIVE-RESIDUE OIL**, and is the type used largely in the commercial packaging of foods.

TASTING THE FRUIT OF THE TREE

> "**L**'huile qu'on tire ici des plus belles
> olives du monde remplace le beurre et
> j'apprehendais bien. Main j'en ai gouté dans
> les sauces et il n'y a rien meilleur." RACINE.

new oil. In cooking it is best to use a good measure of oil and pour off the excess, filtering it and storing it out of the light. All olive oil should be stored, away from heat, light and air. As with all other fats and oils, it must never be allowed to smoke when being heated. This is a sign of overheating; the oil turns black and forms acrolein, which is a toxin.

You may sometimes see olive oil with a white solid layer at the bottom of the bottle. This happens when it has been exposed to the cold, and will not damage or change the oil in any way. When returned to normal room temperature, the oil becomes liquid again. Its solidification point is 36°F. So don't store olive oil in a refrigerator.

L'Olivier Shop, Rue de Rivoli, Paris

OLIVE OIL
AND HEALTH

"Olive oil makes all your aches and pains go away."

ANCIENT PROVERB

*F*or centuries the benefits of olive oil nutritionally, cosmetically and medicinally have been recognized by the people of the Mediterranean. In the Bible, for instance, its healing powers are amply demonstrated in the famous parable, where the Good Samaritan tends to the robbed man by pouring oil and wine into his wounds.

In Greek and Roman times, people cleaned their skins by rubbing themselves with olive oil, then scraping it off with a curved blade of wood or bronze called a strigil. Housed in the British Museum today, there is a fine example of a bronze pot for holding olive oil with two strigils, as used by Greek athletes. Olive oil was also used to maintain the suppleness of skin and muscle, to heal abrasions and to soothe the burning and drying effects of sun and wind. Women used it especially to give body and shine to their hair. Mixed with spices or herbs, it was administered both internally and externally for health and beauty.

Both Pliny and Hippocrates prescribed medications containing olive oil and olive leaves, as cures for a number of disorders ranging from inflammation of the gums, insomnia, and nausea to boils. Many of these old remedies have passed into folk medicine and are still as relevant today as they were hundreds of years ago.

Byron's description of the Mediterranean as "where the citron and the olive are the fairest of fruit" was apt. The influence of both is to be found in the cooking of Spain, the South of France, Italy, Greece, Turkey, Morocco, Tunisia, Portugal, Egypt, Jordan, Lebanon and all the islands which are scattered between these shores. The ingredients which go into Mediterranean dishes are the products of unsophisticated, agricultural communities. In this harsh and arid landscape, only hardy and vigorous trees can survive. Nowhere do you find the lush green grass of northern

FOLK REMEDIES USING OLIVE OIL

FOR SHINING HAIR. After shampooing rub in a mixture of olive oil and egg yolk, juice of a lemon and a little beer. Leave for 5 minutes and wash out.

TO PREVENT DANDRUFF. Rub into the hair a mixture of olive oil and eau de Cologne. Then rinse.

FOR DRY SKIN. Make a face mask with an avocado and olive oil. Leave for 10 minutes and then rinse off.

TO PREVENT WRINKLES. Rub into the skin a mixture of olive oil and the juice of a lemon before going to bed.

TO SOFTEN THE SKIN. Mix together equal parts of olive oil and salt. Massage the body and wash off.

FOR WEAK NAILS. Soak the nails for 5 minutes in warm olive oil and then paint the nails with white iodine.

FOR TIRED FEET. Massage with olive oil.

FOR ACHING MUSCLES. Massage with a mixture of olive oil and rosemary.

TO CLEAR ACNE. Rub with a mixture of 8 fl oz olive oil and 10 drops of lavender oil.

TO REDUCE THE EFFECTS OF ALCOHOL. Take a couple of spoonfuls of olive oil before drinking.

FOR HIGH BLOOD PRESSURE. Boil 24 olive leaves in 8 fl oz of water for 15 minutes. Drink the liquid morning and night for two weeks.

Europe, where cattle can graze, and hence where meat and dairy products are prevalent. What you do find is the olive tree, the vine, lemon and orange trees, wild aromatic herbs, and an abundance of seafood.

Recent research has now provided firm proof that a Mediterranean-style diet, which includes olive oil, is not only generally healthy, but that consuming olive oil can actually reduce cholesterol levels.

The increase in heart disease since the war was an alarming indicator that something in the contemporary industrialized lifestyle was to blame. This led the American Heart Foundation to initiate research into the modern diet, smoking, obesity and high blood pressure. They found that in Greece and especially on the

They also export to the Japanese, who are now one of the most enthusiastic and knowledgeable markets, the United States, Canada, Australia and New Zealand as well as the rest of the EEC.

Siuarana is the second DO and comprises a wide band of country which crosses the Catalonian province of Tarragona from west to east. Within the DO there are two clearly defined districts; the first, further inland, is in the foothills of the Sierra Montsant mountains. Hilly, rugged land — and difficult to cultivate. Nearer the coast lies the Campo de Tarragona, a much smoother terrain with soils of better composition. This DO has around 25,000 acres of olive groves and produces about 2,300 tons of oil per year. The oil from this area is made exclusively from the Arbequina, Rojal and Morrut (Morruda) varieties and has a fine aroma. As with Borjas Blancas, there are two types of oil depending on the ripeness of the olives.

Sierra de Segura has a provisional DO. The olive-growing area is in the north-east of the province of Jaen in Andalucia and covers 94,000 acres of rugged terrain with steep slopes which makes mechanization difficult, so the olives have to be harvested by hand.

The virgin oils of this district are fruity, aromatic and slightly bitter and are mostly made from the Picual olive. This region is truly the heart of Spanish olive growing; I drove for hours round Jaen and saw nothing but rows of olive trees stretching to the horizon. In the early morning during the harvest, landrovers and cars packed inside with pickers and outside with nets and poles headed out of tiny hamlets, towards the fields. In the evenings the processions return, while tractors bearing the day's load of olives make for the local mill. At dusk the drivers stand around patiently smoking and chatting while they wait their turn to drive in, and

Seville

have their cargo carefully weighed and pressed.

The production area of Baena is also provisional, and it extends along the south of the province of Cordoba, between the lowlands and the areas neighboring the Penibetic mountains. Groves of Picudo, Carrasquena, Picual, Hojiblanca and Lechin olives mantle 79,000 acres. The oils from this DO are yellow, with a green and violet tinge and the flavors vary between intense and fruity, and smooth and sweet.

In Spain, as in all the olive-growing countries, there are thousands of small growers who take their olives to the local cooperative where the olives are milled and pressed. The virgin oil produced may then be purchased, bottled and marketed internationally by large distributors such as Ybarra in Seville and Carbonell near Cordoba in the heart of Andalusia who have been producing olive oil since 1866. Carbonell's bottles of oil carry the distinctive and charming picture of a Spanish lady wrapped in a red shawl plucking an olive from a tree.

Extra virgin oils from these companies are excellent, which is not surprising. Because of their size they have the choice of the finest virgin oils in the region. Tankers from local cooperatives deliver to the Carbonell and Ybarra factories, samples are immediately taken from the consignment and then rushed to the laboratory where they are tested for acidity before being accepted. If accepted, the virgin oil may be bottled as extra virgin, depending on the acidity, or added to refined oil and sold as pure olive oil.

As well as being the major producer of olive oil, Spain is justly famous for its table olives, most of which come from Andalusia. The best known of the varieties is the Manzanilla. Picked when green, this is the one you are most likely to buy under the general label "Spanish olives." The Gordal, or Queen as it's also known, the largest and most fleshy of the Spanish olives, is picked both when green and black. Hojiblanca is less highly regarded, but it is a hardy, large cropping variety, and is used for oil as well as being preserved. Then there are Blanguetas, most often marketed at pink-brown ripeness, for eating.

Many of the olives are sold stuffed with pimiento, almonds, onion or garlic. In days gone by these were pitted and stuffed by hand — a labor-intensive job if ever there was one — but today the inevitable machines grade the table olives according to their size, pit them and stuff them at the same rate of 2,000 per minute; what used to take 20 women 8 hours, can now be done in 1 hour by machine.

Recipes using olives and olive oil abound in Spanish cooking. Olive oil is used for frying fish and meat as well as sweet dishes like Churros — deep-fried strips of batter coated in sugar, which the Spanish eat in the morning with hot chocolate or coffee.

It is also used in making dough for tortas and empanadas, and numerous different types of bread. It is added raw as a condiment to improve the texture, taste and nutritive values of dishes such as Gazpacho — and, of course, it appears in green salads and in dishes like hervido, which contains boiled potatoes, zucchini and eggplant, tossed in oil, like a hot salad. Throughout Spain olive oil preserves cheeses, sausages, fish or vegetables, often with the addition of spices or herbs.

Spanish cooking is still unfortunately much underrated, perhaps because Spanish restaurants abroad are still a rarity. I am happy to report, however, that now there are tapas bars springing up in Britain and the U.S. The name comes from the days when a small plate or tapa was placed on a glass of wine or beer to keep out the dust and flies. Then no doubt, one day long ago, an enterprising bar owner popped a couple of olives or anchovies onto the plate as an appetizer and one of the most delightful Spanish traditions was born.

Spain used to be a very poor country but today in cities such as Madrid you witness sophistication and wealth — a stylish younger generation confidently parading its designer labels on the wide elegant streets. Suddenly Spain is drawing the attention of the rest of the world; all things Spanish are now very much in vogue and it could be that the day for Spanish cuisine has also finally come.

AJO BLANCHO DE ALMENDRAS
(WHITE GARLIC SOUP)

A soup made of bread, almonds, oil, garlic and water may sound an unlikely combination but to my mind this is one of the most delicious dishes to serve as a summer appetizer. Friends who have tasted this have all marveled at the ingredients and have usually asked for second helpings. This is a typical Spanish dish dating back to Roman times, when it was known as "what a mess," which meant it contained everything barring the kitchen sink! Usually garnished with white grapes, this version, which has raisins and diced apple, was served to me in the El Caballo Rojo restaurant in Cordoba and it is the best I have tasted.

To serve 4

4 slices stale white bread, without
   crusts
cold water
4 cloves of garlic
1 cup almonds, blanched and
   peeled
4 tablespoons olive oil
3 tablespoons white wine vinegar,
   ideally sherry vinegar

salt, to taste

FOR GARNISH
1 apple, diced
raisins

Soak the bread in water and squeeze dry. Put the salt and the peeled cloves of garlic in a mortar or food processor and grind thoroughly. Add the almonds little by little until everything is mixed to a paste. Add the bread, then start adding a dash of oil drop by drop until you have the consistency of mayonnaise. Add the vinegar drop by drop, beating the whole time. Add enough cold water to obtain a thick beverage. Add more salt and vinegar if necessary. You may sieve the mixture at this point if you wish but I think it is better left unsieved. Chill the soup and just before serving add the diced apple and raisins. You can use washed white grapes instead of the apple and raisins.

N.B. *The proportions of garlic, olive oil and vinegar you use depend entirely on your preference, so taste all the time while you are making this soup.*

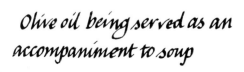

Olive oil being served as an accompaniment to soup

ENSALADA DE NARANJAS, ACEITUNAS Y JEREZ

(ORANGE, OLIVE AND SHERRY SALAD)

 There is a dazzling array of fruit grown in Spain. The market stalls all over the country display kaleidoscopic arrangements of figs, melons, lemons, grapes, apples, dates, strawberries and of course oranges, which are used in this refreshing salad.

To serve 4

4 large oranges
$\frac{1}{2}$ cup olive oil
$\frac{1}{4}$ cup dry sherry
1 tablespoon fresh mint, chopped
$\frac{1}{3}$ cup black olives

3 tablespoons raisins
salt and pepper, to taste

FOR GARNISH
sprigs of fresh mint

Thinly pare the rind from one orange and cut the peel into matchstick strips. Squeeze the juice from the orange. Mix the olive oil with the orange juice, sherry, strips of orange peel, mint, salt and pepper. Peel the remaining oranges and cut into slices. Arrange the orange slices in overlapping rings on a serving dish. Scatter the olives and raisins over the oranges and spoon over about two-thirds of the dressing. Cover and chill for about 2 hours. Just before serving spoon over the remaining dressing and garnish with the sprigs of mint.

ALL-I-OLI
(CATALAN OLIVE OIL AND GARLIC SAUCE)

 This classic Catalan sauce is not a mayonnaise because it does not contain eggs, just olive oil and garlic, which makes it a tricky sauce to make since it has a tendency to separate. The answer is to add a small quantity of white crustless bread if it will not thicken. The number of cloves of garlic used will depend on your personal taste and you can add lemon juice if you wish. It is traditionally served with grilled meat or fish or mixed with Romesco.

Makes 1 cup

6–10 cloves of garlic, peeled   salt, to taste
1 cup olive oil

Put the cloves of garlic in a blender with the salt and blend at high speed. With the motor running add the olive oil very slowly until the sauce begins to thicken. If it remains oily add the bread.

ROMESCO

 Catalonian cooking is an assimilation of many influences including French and Italian and, as such, sauces form the basis of many dishes. Romesco is the great Catalan sauce. It takes its name from the hot red romesco peppers which grow in this region. The fieriness of the sauce can vary from the tolerable to the explosive, so if you come across it being served in Spain, treat it with respect. In Catalonia, Romesco is often served with a bowl of all-i-oli and the two are mixed to taste at the table. The real romesco peppers, which may prove difficult to find outside Spain, can be substituted with chilli peppers but, of course, the taste will not be authentic. The sauce is used to accompany shellfish, fish, meat, poultry as well as stews and salads.

To make about $1\frac{1}{4}$ cups

$\frac{1}{2}$ cup blanched almonds $\frac{2}{3}$ cup olive oil
3 cloves of garlic, unpeeled   3 tablespoons red wine vinegar
2 tomatoes, whole   salt, to taste
1 dried romesco pepper or 1 dried hot
  chilli pepper

Preheat the oven to 350°F. Place almonds, garlic, tomatoes and pepper on a baking tray and put in the oven for 10–15 minutes, removing the almonds when they are lightly browned and the tomatoes and garlic when they are soft. Peel the tomatoes and remove the seeds, peel the garlic. Place the tomatoes, garlic, pepper and almonds in a blender and grind thoroughly. Add the vinegar and salt and then gradually blend in the oil, until the mixture thickens.

SALSA DE ACEITUNAS
(GREEN OLIVE SAUCE)

 A simple and quick sauce which can be served with grilled meat or fish or also over pasta.

Makes about $1\frac{1}{4}$ cups

$\frac{2}{3}$ cup green olives, pitted
1 small onion, peeled and sliced
2 cloves of garlic, peeled
2 tomatoes, peeled, seeded and chopped

$\frac{1}{4}$ cup white wine or sherry
1 tablespoon lemon juice
3 tablespoons olive oil
$\frac{1}{2}$ cup water
salt and pepper, to taste

Place the olives, onion, garlic, tomatoes, wine and lemon juice in a food processor and blend until smooth. Add the olive oil slowly and salt and pepper if necessary. Put the mixture into a saucepan with the water and simmer for 10 minutes.

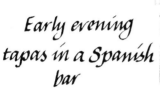

Early evening tapas in a Spanish bar

BONITO CON ACEITUNAS
(TUNA WITH OLIVES)

 Andalusia is renowned throughout Spain for the lightness of its fried food, especially fried fish. In such a hot climate there is no place for heavy dishes and this recipe is typical of the simple but imaginative way much Andalusian food is prepared.

To serve 4

4 tuna steaks
$\frac{2}{3}$ cup olive oil
$\frac{3}{4}$ cup white wine
1 teaspoon wine vinegar
2 cloves of garlic, peeled and
   crushed

1 bay leaf
sprig of thyme
flour
$\frac{2}{3}$ cup green olives, pitted and
   chopped
salt and pepper, to taste

Place tuna steaks in a shallow dish. Make the marinade with the olive oil, wine, vinegar, herbs, garlic and salt and pour over the fish. Leave to stand in a cool place for at least 2 hours. Remove the fish from the marinade, pat dry and dip into the flour, shaking off the excess. Heat some olive oil in a deep pan and fry the steaks until golden on both sides and the flesh is tender. Remove and keep warm. In the same pan add the marinade and boil until it has reduced a little. Strain it through a sieve, add the chopped olives and pour over the fish.

CHURROS
(FRIED PASTRIES)

 Churros are fluted, sausage-shaped, deep-fried fritters which the Spanish eat warm, dusted with sugar, dipped into their milky coffee or chocolate, for breakfast and through the morning.

To serve 4

2 cups water
$\frac{1}{2}$ teaspoon salt
2 cups flour, sieved

olive oil, for deep frying
sugar, to taste

Bring the water and salt to the boil. Remove the pan from the heat and pour in all the flour. Beat the mixture well, until it pulls away from the sides of the pan and forms a mass. Heat the olive oil in a pan until it is very hot, about 350°F on a deep-fat thermometer, or until a small cube of stale bread dropped into the oil turns golden in 1 minute. Put the paste into a piping bag fitted with a star nozzle and pipe 8″ lengths into the oil, cutting them with a knife or scissors. Only do a few lengths at a time and fry for 5–8 minutes or until they are crisp and golden brown. Using a slotted spoon remove the churros to paper towel to drain. Sprinkle them with sugar and serve while they are still warm.

TORTAS DE ACEITE
(SESAME SEED AND ANISEED COOKIES)

 Most people probably only think of using olive oil in savory dishes but it is so versatile it can equally well be used in sweet dishes as well. These spicy little cookies are perfect with coffee.

To make 24 cookies

$1\frac{1}{2}$ cups olive oil
thinly pared rind of $\frac{1}{2}$ lemon
1 tablespoon sesame seeds
1 tablespoon aniseeds
$\frac{1}{2}$ cup dry white wine
2 teaspoons grated lemon rind
2 teaspoons grated orange rind

$\frac{1}{2}$ cup sugar
5 cups flour
1 teaspoon ground cinnamon
1 teaspoon ground cloves
1 teaspoon ground ginger
$\frac{1}{3}$ cup flaked almonds

Heat the olive oil in a saucepan until a light haze forms above it. Remove the pan from the heat and add the pared lemon rind, sesame seeds and aniseeds and leave to cool. Remove the lemon rind and pour the oil mixture into a large bowl. Add the wine, lemon and orange rind and sugar and beat until well blended and the sugar has dissolved. Sift the flour and spices into the bowl a little at a time, beating well until they form a soft dough. Using your hands, lightly knead the dough until it is smooth. Form into a ball and wrap in wax paper. Set aside at room temperature for 30 minutes.

Preheat the oven to 400°F. Line two large cookie sheets with non-stick cooking parchment. Remove the dough from its wrapping and divide into 24 equal pieces. Roll the pieces into small balls and using the palm of your hand, flatten them into flat round cookies about $\frac{1}{4}″-\frac{1}{2}″$ thick. Arrange the cookies on the prepared cookie sheets and press a few flaked almonds into the top of each one. Put the sheets into the oven and bake for 15–20 minutes or until the cookies are firm to the touch and golden brown round the edges.

ITALY

*I*taly is synonymous in most people's minds with olive oil. Looking back through history we see there are plenty of good reasons for this. As much a part of the people as the countryside, the olive probably first came to Italy from Sicily in the 6th or 7th century BC, and gradually it flourished in the ideal climate and varied soils of the peninsula. The Ancient Romans were the most influential in spreading the cultivation of the olive tree throughout the rest of Europe and North Africa, during the growth of their empire. They perfected curing techniques for olives and by the invention of the screw press,

PAPPA AL POMODORO
(TOMATO SOUP)

 This is a Tuscan soup, the main ingredients being stale bread and tomatoes. It is a soup of peasant origins and, like so many of these regional Italian dishes, it shows the ingenuity in combining a few basics and creating a simple but delightful meal. It is finished off with a dressing of olive oil.

To serve 4

3 large cloves of garlic, sliced
1½ lb tomatoes, peeled,
   seeded and chopped
8 slices of bread, a few days old

3¾ cups water
large bunch of fresh basil leaves
olive oil
salt and pepper, to taste

Heat some olive oil in a pan, add the garlic and cook gently for about 5 minutes. Add the tomatoes and simmer for 15 minutes. Break the bread into small pieces and add to the pan along with the basil. Cover with water and simmer for 15 minutes. Season with salt and pepper.

   The flavor of the soup improves if left overnight. So it can either be reheated or served cold. At the table pour over some olive oil.

Estate in Tuscany

BAGNA CALDA
(HOT GARLIC AND ANCHOVY SAUCE)

 A classic sauce from Piedmont which is for true garlic lovers only. The name means "hot bath" and it is served in the same way as a fondue, in a pot on the center of the table over a candle, surrounded by raw vegetables, which are dipped into the sauce. It makes a wonderfully convivial meal for a group of friends and should be served with plenty of full-bodied red wine and good crusty bread.

To serve 6–8

$\frac{1}{4}$ cup butter
5 cloves of garlic, finely sliced

8 anchovy fillets in oil, drained
1 cup olive oil

Heat the butter in the serving pot or a small saucepan and add the garlic. Keep on a low heat so that the garlic does not brown. Add the anchovies and stir well, then gradually add the oil. Cook for about 10 minutes, over a low heat, stirring constantly. Place on the table over a candle or spirit lamp. Serve with a selection of washed and trimmed raw vegetables such as sliced peppers, celery, fennel, cauliflower, mushrooms, or carrots.

CAPONATA
(EGGPLANT SALAD)

 A Sicilian dish which can be served as an antipasti. It can also be stored in glass jars which should then be sealed and boiled for 20 minutes, after which it will keep for months.

1lb eggplant, cut into cubes
2 tablespoons olive oil
4 sticks of celery, sliced
1 onion, sliced
8oz tomatoes, peeled and
   chopped

2 tablespoons capers
2 tablespoons pine nuts
1 tablespoon sugar
$\frac{1}{2}$ cup red wine vinegar
$\frac{1}{4}$ cup large green olives
salt and pepper, to taste

Sprinkle the cubes of eggplant with salt and leave them in a colander to drain for an hour. Then dry them thoroughly on a paper towel. Heat plenty of olive oil in a pan and fry the eggplant cubes until they are brown. Drain on paper towel. Fry the celery in the same oil as the eggplant and when brown remove from the pan. Add the onion and cook until soft, add the tomatoes and cook gently for 10 minutes. Add all the other ingredients, return the eggplant and celery to the pan and simmer for a further 5 minutes. Allow it to go cold before serving.

PEPERONI RIPIENI
(PEPPERS STUFFED WITH TOMATOES, TUNA, ANCHOVIES AND OLIVES)

 This is a dish I have been serving for years. It encompasses all that I love about Italy and the stuffing for the peppers is so good I've usually eaten half of it before I start filling the peppers!

To serve 4

4 large red or green peppers
4 tablespoons olive oil
1 onion, sliced
2 cloves of garlic, crushed
$1\frac{1}{2}$ lb fresh tomatoes, peeled
  or canned peeled tomatoes
2 tablespoons tomato puree
a few leaves of fresh basil or
  $\frac{1}{2}$ teaspoon dried basil

$\frac{1}{2}$ teaspoon oregano
1 tablespoon parsley, chopped
12oz can of tuna in olive oil,
  drained
4 anchovy fillets, chopped
12 black olives, pitted and halved
2 teaspoons capers
$\frac{1}{3}$ cup parmesan, grated
salt and pepper, to taste

Preheat the oven to 325°F. Slice the tops from each pepper. Remove and discard the membrane and seeds. Remove the stems from the sliced pepper tops and dice the flesh. Heat the olive oil in a pan and add the onion, garlic and diced pepper. Saute until the onion is soft. Stir in the tomatoes, tomato puree, herbs, salt and pepper and cook for 10 minutes. Stir in the rest of the ingredients except the cheese and cook for a further 10 minutes. Spoon the mixture into the peppers. Pour about $\frac{1}{4}''$ of olive oil into a baking pan and arrange the peppers standing upright on the pan. Bake for about 30 minutes or until the peppers are soft, basting occasionally with the oil in the pan. When the peppers are cooked, sprinkle the parmesan cheese on top of the filling and bake for a further five minutes. Remove from the oven and serve immediately.

PEPERONI ARROSTITI
(PEPPER SALAD)

 This simple salad can be served as part of an antipasti, but I especially like to serve it in the summer to accompany barbequed fish and meat. Its only dressing is olive oil so use a fruity one.

To serve 4–6

2 each red, yellow and green peppers

2 cloves of garlic, chopped
olive oil

Blacken the skins of the peppers by holding them over a gas flame or placing under the broiler. Remove the skins and wash the peppers. Cut into wide strips and remove the seeds and membrane. Sprinkle with garlic and cover with olive oil.

FOCCACIA
(FLAT BREAD)

Every region in Italy has a different version of this delicious bread which is like a pizza dough. It can be flavored with sage, crushed olives and onions or topped like a conventional pizza with tomatoes and cheese. This is the plain version just flavored with olive oil and dimpled on the top. The aroma of the dough and the fruity oil when it is cooking makes it difficult to leave it long enough to cool! It is excellent with cheese or served with antipasti.

To serve 4–6

1oz fresh yeast
$\frac{2}{3}$ cup warm water
4 cups flour

6 tablespoons olive oil
salt, to taste

Dissolve the yeast in the warm water. Put the flour into a large bowl and pour in the yeast, salt and 3 tablespoons of the olive oil. Knead these together until you have a smooth, soft, pliable dough. Cover, and leave to rise in a warm place until the dough has about doubled in size. Knock back the dough and knead for a few minutes. Roll out the dough to about $\frac{1}{2}''$ thick and lay it on a well oiled rectangular baking pan. Cover with a damp cloth and allow it to rise again for about 30 minutes. Meanwhile preheat the oven to 400°F. Before putting into the oven, dimple the top by pressing the dough with your fingertips and pour over the remaining olive oil. Bake for 20–30 minutes.

FRANCE

"*P*rovence is a painter's paradise and its tree, the olive is the painter's tree." Aldous Huxley's simple but evocative description of the heart of France's olive-growing region, points to the great beauty of the countryside — the strong, sunny colors and the cornucopia of good things which spring from the soil. The hills are clad in wild fragrant rosemary, thyme and fennel. Fields of

The big "freezes" of 1929 and 1956 affected the trees badly. Moreover, the olive trees in the South of France have been cut down to make way for the now more profitable vine.

The olive industry in France consists entirely of some 45,000 small producers, who take their olives to local mills or co-operatives. There are no Bertollis or Carbonells to buy up the crops, and market them on an international scale. This is one reason why you do not find that many French olive oils outside France. The small farmer just doesn't have the resources to advertise and ship his product overseas. The competition from seed oils, at much lower prices, has also been an added difficulty for the small producer and has encouraged the French housewife to abandon olive oil in favor of sunflower and peanut oil.

Even if the French housewife does not hold olive oil in high regard, gourmets do and claim that there are two elements in Provencal cooking, olives and garlic — which supports the Provencal saying "a fish lives in water and dies in olive oil"!

When you travel through Provence you will taste so many traditional dishes, in which olive oil plays a major role. Dishes like Bouillabaisse, a rich fish stew which purists claim cannot be made outside of the South of France. This is because rascasse, a tiny rock fish and the essential ingredient, giving the dish its unique flavor, is only found along these coastal waters. Peppers, zucchini and eggplant grown locally in abundance appear in Ratatouille, the vegetable stew. Touiller in Provencal means to stir, and tatouille, to stir a second time, which is exactly what you have to do when preparing this dish. The French in this part of the country even manage to create their own form of butter from olive oil, by putting a cup of oil in the fridge and letting it solidify. Soups are prevalent and one famous one, L'Aigo Boulido, is garlic boiled in water and poured over bread soaked in olive oil. It may not sound very appetizing but it is supposed to be a marvelous cure for a hangover!

Olive oil and soap shop. Marseilles

SOUPE AU PISTOU
(VEGETABLE SOUP WITH PISTOU)

 I love making soups. I started making them in my student days because vegetable soups are filling, cheap and nourishing — a pound of carrots, some leeks and maybe some fresh herbs, if they are available, and you have a meal in half an hour! For this reason, it's a constant source of amazement why anyone buys ready-made soups even on the basis of convenience. My repertoire of soups for both summer and winter has grown over the years and this is one of my personal favorites. A friend of mine has always categorized my soups as queens, kings and aces. This is an ace!

To serve 4 (or in my case 2 with second helpings)

$\frac{2}{3}$ cup dried white kidney
  beans
water for soaking and boiling
  beans
1 onion, chopped
1 clove of garlic, crushed
$1\frac{1}{4}$ cups tomatoes, peeled,
  seeded and chopped
$\frac{2}{3}$ cup carrots, diced
$\frac{2}{3}$ cup potatoes, diced
2 leeks, sliced
$\frac{1}{2}$ cup celery, sliced
$\frac{3}{4}$ cup green beans
1 cup zucchini, sliced

$\frac{1}{2}$ cup either broken spaghetti
  or small pasta for soups
$6\frac{1}{4}$ cups water
2 tablespoons olive oil
salt and pepper, to taste

PISTOU
3 cloves of garlic
large bunch of basil leaves
$\frac{2}{3}$ cup parmesan, freshly grated
2 tablespoons olive oil

Put the beans in a pan with plenty of cold water. Bring to the boil and then remove from the heat. Leave the beans to soak for 1 hour. Then cook until tender. Heat some olive oil in a large pan and cook the onion and garlic until golden brown. Add the tomatoes and cook for a few minutes. Pour in the water and bring to the boil. Add the carrots, potatoes, leeks and celery. Reduce the heat and simmer for 10–15 minutes. Meanwhile make the pistou. You can use a mortar and pestle or a blender. Mash the garlic and basil together. Add the cheese and then gradually beat in the olive oil. Put in the beans and their liquid, the green beans, zucchini and pasta into the soup. Simmer until tender. Season to taste. Put the pistou into a soup tureen and pour the soup over it. Leave the soup to stand for about 10 minutes to allow the pistou to flavor it. Pistou should never be added to the soup while it is cooking and should be regarded as a condiment to be added as the final flourish.

AIGO BOULIDO
(BOILED WATER)

 This Provencal soup is supposed to be a panacea for all ills and a marvelous cure for hangovers. There is an old saying "Aigo Boulido sauova la vida". Boiled water saves your life!

To serve 4

4 cups water
12 cloves of garlic
2 bay leaves
1 sprig of sage

4 tablespoons olive oil
salt, to taste
slices of dry white bread

Put all the ingredients, except the bread and 1 tablespoon of olive oil, into a saucepan and boil for 15 minutes. Remove the pan from the heat and let it stand for 5 minutes. Put slices of dry bread in the bottom of each soup plate, dribble over the remaining oil and pour over the soup.

SALAD NICOISE

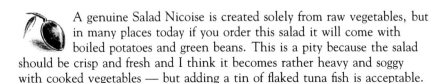 A genuine Salad Nicoise is created solely from raw vegetables, but in many places today if you order this salad it will come with boiled potatoes and green beans. This is a pity because the salad should be crisp and fresh and I think it becomes rather heavy and soggy with cooked vegetables — but adding a tin of flaked tuna fish is acceptable.

To serve 6 as an appetizer or 4 as a main dish

10 medium-sized tomatoes, cut
　　into quarters
1 clove of garlic, peeled
1 cucumber, peeled and finely
　　sliced
2 green peppers, membrane and seeds
　　removed, thinly sliced

6 scallions, thinly sliced
12 anchovy fillets
1 cup black olives
3 hard boiled eggs, quartered
6 tablespoons olive oil
salt and pepper, to taste
a few shredded basil leaves

Sprinkle the tomatoes with salt and set on one side to drain. Cut the clove of garlic in half and rub a salad bowl well with the two halves. Make the dressing with the olive oil, salt, pepper and basil. Add all the ingredients including the drained tomatoes to a bowl and pour over the dressing.

RATATOUILLE

Even though all the vegetables that go into a ratatouille are available now, all the year round, I still think of this dish as heralding in the summer. It certainly contains the essence of the Mediterranean. I include it because it is inconceivable that it would taste as good as it does without olive oil. It can be served hot or cold, as a main or side dish.

To serve 4 as a main dish or 6 as an appetizer

2lb eggplant, cut into cubes
1½ lb zucchini, sliced
2 large green peppers, membrane and
    seeds removed, cut into squares
1 large onion, sliced

2 large cloves of garlic, crushed
2lb tomatoes, peeled and
    seeds removed
olive oil
salt and pepper, to taste

Sprinkle the eggplant with salt and leave to drain for 10 minutes, then pat dry. Heat a good measure of olive oil in a pan and fry the eggplant quickly, so as to prevent it from soaking up too much oil, until it is light brown. Remove it from the pan. Repeat with the zucchini, peppers, onion and garlic, one after the other, adding more olive oil to the pan as necessary. When all the vegetables have been browned, return them to the pan and gently stir in the tomatoes, so they coat all the vegetables. Bring to the boil and then simmer gently until the vegetables are tender. Season to taste. Do not overcook them or you'll end up with a soggy stew. Each vegetable should hold its shape.

POMMES A L'HUILE

This is a wonderful, hot potato salad, to serve with fish, or the way I usually serve it, with boiled sausage. A dish in which you can use your fruitiest olive oil.

To serve 4

1lb new potatoes
handful of parsley, chopped
2 scallions, sliced

3 tablespoons olive oil
salt, to taste

Boil the potatoes in their skins, in salted water. When they are cooked, drain and put into a serving bowl. Stir in the parsley, scallions and olive oil.

PAN BAGNAT

The olive growing regions of France follow a line round the south coast so it's not surprising to find that many recipes using olives and olive oil come from in and around Nice. This snack is really a salad nicoise in bread; pan bagnat means literally wet bread and this is ideal to take on a summer picnic.

1 long French loaf or individual
  rolls
1 clove of garlic
olive oil
wine vinegar, to taste
1lb tomatoes, sliced
1 onion, sliced

1 green pepper, seeded and
  sliced
1 red pepper, seeded and sliced
1 hard boiled egg, sliced
2–3 anchovy fillets
6 black olives, pitted
salt and pepper, to taste

Slice the loaf or rolls horizontally and scoop out a little of the bread to allow the filling to fit. Rub the inside of each half with the garlic, pour on a few drops of the vinegar and plenty of olive oil, salt and pepper. Fill the loaf with the tomatoes, onion, peppers, egg and anchovies. Press the top of the loaf on like a lid. Wrap in aluminum foil and put under a heavy weight for about 1 hour to allow the juices to seep into the bread. Slice and serve.

PISSALADIERE

 This dish is a native of Nice and is similar to an Italian pizza. It is a bread dough base covered with a generous layer of onions, olives and anchovies. It can be made equally well on a shortcrust pastry base and this is the version I give here.

8oz shortcrust pastry
$2\frac{1}{2}$ lb onions, finely sliced
2 cloves of garlic, crushed

10 anchovy fillets
12 black olives
2 tablespoons olive oil

Preheat the oven to 400°F. Roll out the pastry and line a 12″ oiled baking tray. Heat the olive oil in a pan, add the onions and garlic and cook very slowly until the onions are soft but not brown. Top the pastry with the cooked onions, decorate with the olives and anchovies. Drizzle a little olive oil over the top and bake in the oven for about 20 minutes or until the pastry is cooked. Serve hot or warm.

MAYONNAISE

 It may seem rather obvious to have a recipe for mayonnaise, but I think it is worth including because it has to be the quintessential olive oil sauce. I personally find that the finest extra virgin oils are too flavorful to use in mayonnaise — they make the taste too rich — and so I usually opt for a good brand of pure olive oil. Pure olive oil is refined olive oil with some extra virgin added for taste, aroma and color. So its flavor is not too powerful.

2 egg yolks
$1\frac{1}{2}$ tablespoons white wine vinegar or
    lemon juice

$\frac{1}{2}$ teaspoon mustard
$1\frac{1}{4}$ cups olive oil
salt and pepper, to taste

Blend together the egg yolks, mustard and vinegar or lemon juice in a food processor, then with the motor running add the olive oil very slowly in a thin stream. If the mayonnaise becomes too thick, thin it with either more vinegar and lemon juice or some warm water. The whole process is more assured of success if everything is at room temperature.

    With a plain, homemade mayonnaise you can add all sorts of other ingredients, such as tomato puree, horseradish, anchovies, fresh herbs — basil, tarragon, dill, chervil or watercress for a beautiful green mayonnaise, which can be served with broiled fish, chicken or raw vegetables.

The importance of the olive varies from region to region. In Crete, olive trees occupy over 60% of the cultivated land, in the southern mainland, 30%, but in the north only 4%. Production is concentrated in the Peloponnese and Crete. One of the main towns in the Peloponnese is Kalamata, which is of course also the name of the main olive variety grown in Greece.

Many families rely heavily on olives and olive oil for their income. There are nearly 1 million farms in Greece and the greater part of these are engaged in olive growing. But the olive industry doesn't only employ a vast labor force for its cultivation; there are people working in the 3,000 traditional and 400 modern mills, as well as the bottling and packing plants. There is no other crop in Greece which occupies so much available workforce.

Olive cultivation is highly seasonal work; November to February is the harvest time and fits in ideally for the Greeks with the tourist season in the summer months. This means that families, who would have difficulty surviving on olive growing alone, can work on casual summer jobs or perhaps rent out rooms to the thousands of sun-seeking vacationers.

The market in Athens displays an amazing array of green olives, cracked and preserved in rigani, lemon and coriander; and black olives, especially the large juicy kalamatas. Eighty per cent of the olives grown go to produce oil, and the rest are preserved as table olives. Europe and the United States have been enthusiastically devouring Greek olives for years but the olive oils have, as yet, not found the same favor. They are an acquired taste, being much more powerful and rustic than the oils of Spain or France but a perfect foil to retsina. Maybe this is why the Greeks resinate their wine.

Olives are not used generally in Greek dishes, but they are eaten in abundance as a part of mezes and it is unusual to be served a drink without a dish of large, shining olives to accompany it.

I can't think of anything more enjoyable than sitting in a taverna on a warm summer's evening with a chilled bottle of retsina, some grilled fish which has been cooked over an open fire, and a plate of green beans bathed in a pool of fruity olive oil, waiting to be mopped up with lashings of fresh bread.

FAKI SOUPA
(LENTIL SOUP)

 A few years ago when I was staying on the tiny island of Astipalaia, I was delighted to find this soup on the menu of one of the little tavernas. I love lentil soup but I had not come across it in Greece before. The owner, impressed by my enthusiasm for her cooking, gave me her recipe.

To serve 4

6oz brown lentils	a good pinch of oregano
1 large onion, sliced	2 tablespoons olive oil
2 cloves of garlic, crushed	3 cups water
3oz can of tomato puree	salt and pepper, to taste
8oz fresh tomatoes, peeled and seeded	

Cover the lentils with cold water and bring to the boil. Drain the lentils and return to the pan with 3 cups of water, the garlic, onion, tomato puree, fresh tomatoes, oregano and olive oil. Bring to the boil and simmer until the lentils are soft. Remove from the heat and blend but not for long; the mixture should not be too smooth. Season to taste.

TARAMASALATA

 Tarama is dried, salted, gray mullet roe but as it is not easily available, most people use smoked cod's roe. Like all these traditional dishes there are fierce debates about authentic recipes — some will say that it should not contain garlic, some recipes contain onion and so on. This is my preferred way.

4oz smoked cod's roe	$\frac{1}{2}$ small onion, sliced
2 slices of white bread, with crusts removed	juice of a large lemon
cold water	$\frac{1}{2}$ cup olive oil
	black olives, to garnish

Soak the bread in cold water for 5 minutes, then squeeze dry. Skin the cod's roe and put in a blender, with the bread, onion and half the lemon juice and process. Gradually add the oil until you have a creamy paste. Add the rest of the lemon juice according to your taste. Garnish with some black olives and serve with warm pita bread.

MELIZANOSALATA
(EGGPLANT SALAD)

 This puree of eggplant is known in some places as "poor man's caviar" and can be served as part of a mezes, the little portions of hot and cold food served with drinks. If you like eggplant this is a delicious way to serve it and it tastes even better if baked on charcoal, so have a go at putting it on the fire the next time you have a barbeque.

To serve 4

1lb eggplant
1 small onion, sliced
1 clove of garlic, crushed
juice of 1 lemon

$\frac{1}{4}$ cup olive oil
parsley, chopped
black olives, to garnish

Preheat the oven to 400°F. Prick the eggplant, put it on a baking tray and place in the oven until it is soft — about 1 hour depending on its size. Don't worry if the skins turn black as this gives a lovely smokey flavor to the dish. When it is cool enough to handle, scoop out the flesh and put into a blender with the garlic, onion and half the lemon juice. Blend well and add the oil until you have a thick paste. Add the rest of the lemon juice according to taste. Serve in a dish, garnished with the chopped parsley and olives.

SKORTHALIA
(GARLIC SAUCE)

 This pungent garlic sauce is served in Greece with fried fish or vegetables. I have had it as an accompaniment with fried salt cod as an appetizer and found it compulsive eating.

3 slices of white bread, crusts
   removed
cold water
3 large cloves of garlic

juice of $\frac{1}{2}$ lemon or
  2 tablespoons
  white wine
  vinegar
$\frac{1}{2}$ cup olive oil

Soak the bread in cold water for 5 minutes and squeeze dry. Put into a blender with the garlic,

lemon juice or wine vinegar and process well. Gradually pour in the olive oil until you have a sauce the consistency of mayonnaise.

VEGETABLES A LA GRECQUE

Some people describe the vegetable dishes from Greece as swimming in olive oil and that is exactly how they are meant to be — it is what makes them so good eaten cold. They can be cooked with a variety of aromatics but the essential ingredient is good Greek olive oil. This is a marinade I use to cook a selection of vegetables, which I serve as an appetizer with plenty of crusty bread to mop up the juices.

To serve 4–6

MARINADE
$2\frac{1}{2}$ cups water
$2\frac{1}{2}$ cups dry white wine
$\frac{2}{3}$ cup olive oil
juice of one lemon
2 cloves of garlic, crushed
2 bay leaves
$\frac{1}{2}$ teaspoon coriander seeds
1 sprig of thyme
5 peppercorns
small handful of parsley sprigs

VEGETABLES
a selection from the following:
    small artichokes, quartered;
    button mushrooms; cauliflower;
    bulb fennel; small pickling
    onions

Bring the wine and water to the boil in a large pan and add all the other marinade ingredients. Reduce the heat and simmer for 15 minutes. Then throw in the vegetables, putting the longer cooking vegetables in first. Cook for about 10 minutes and then add the mushrooms and cauliflower and cook for about a further 5 minutes. Strain the liquid off from the vegetables, and boil to reduce to about half. Pour over the vegetables and refrigerate. Serve cold.

HORIATIKI
(GREEK COUNTRY SALAD)

Until a few years ago, I went to Greece seven years in succession. Though there are many Greek dishes I savor, this is the one I most associate with my many happy vacations.

To serve 4

1lb tomatoes, cut into quarters
½ cucumber, peeled and sliced
1 small green pepper, membrane and seeds removed, thinly sliced
1 onion, thinly sliced

5oz feta cheese
12 black olives
a pinch of oregano
4 tablespoons olive oil
salt and pepper, to taste

Place the tomatoes, cucumber, pepper and onion in a bowl. Dress with the olive oil, salt and pepper. Arrange the feta cheese and olives on the top and sprinkle on the oregano.

KOTOPOULO ME ELIES
(CHICKEN CASSEROLE WITH OLIVES)

This spicy casserole is based on one from Rena Salaman's excellent book *Greek Food*. She recommends using a boiling fowl as its long cooking time gives the spices time to flavor the meat.

To serve 4

1 boiling fowl or a jointed chicken
3 tablespoons olive oil
⅔ cup red wine
1 stick of cinnamon
3 cloves

2 allspice
14oz can of tomatoes
16 black or green olives
1 bay leaf
salt and pepper, to taste

Wash and dry the chicken, season with salt and pepper and fry in the olive oil. Pour the wine into the pan, add the spices, bay leaf and tomatoes. Cover and simmer for 1 hour for chicken and about 2 hours for a fowl or until the meat is tender. Add the olives in the last 10 minutes, by which time the sauce should be thick. Serve with rice or pasta.

NORTH AFRICA

Despite the particular difficulties of growing olives in North Africa, Morocco, Tunisia and Algeria excel in their table olives.

Olives are picked at every stage of ripeness and the markets are resplendent with enormous bowls of green olives with pimiento and preserved lemons, pink olives, red olives, brown olives, and black olives, colors you never imagined olives could be. They are preserved in

Moroccan market

HARISSA
(HOT CHILLI SAUCE)

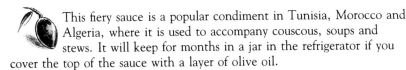

This fiery sauce is a popular condiment in Tunisia, Morocco and Algeria, where it is used to accompany couscous, soups and stews. It will keep for months in a jar in the refrigerator if you cover the top of the sauce with a layer of olive oil.

4oz dried, red chilli peppers
4 cloves of garlic

$\frac{2}{3}$ cup olive oil
salt, to taste

Split the chillis and remove the stems and seeds. Soak the chillis in warm water for about an hour until they are soft. Put them in a mortar or blender, with the garlic and some salt. Pound to a paste, then add the olive oil gradually until you have a smooth sauce.

MOROCCAN ORANGE AND OLIVE SALAD

Oranges and lemons are used quite commonly in salads in Morocco, the lemon often being pickled first for a few weeks. Additions to this delicious and unusual salad can be finely sliced onion and fresh herbs like mint or marjoram. This dish is excellent served with the rich tagines found in Morocco or with roasted meat.

Serves 4

4 large oranges
4oz black olives
4 tablespoons olive oil
2 tablespoons lemon juice

$\frac{1}{2}$ teaspoon Dijon mustard
salt and freshly milled black
pepper

Peel the oranges and remove all the white pulp. Cut into thin slices and reserve any juice. Pit the olives and put the olives and oranges into a bowl. Put all the remaining ingredients including any orange juice into a screw-topped jar and shake until blended. Pour this mixture over the oranges and olives and toss well. Chill slightly before serving.

THE AMERICAS

*T*he Carthaginians, Greeks and Romans introduced the olive into their "New World," and in the same way, when the Spanish missionaries ventured to Paraguay and Peru, in the 16th century, they brought the olive to South America, where it spread into Mexico and finally into California, which is now the main olive-growing area in the U.S.

In world terms, olive oil produced in California is a drop in the ocean, somewhere between 0.4–0.5%! Moreover, the

OLIVE OIL IN THE STORES

During the research and writing of this book, I have sampled nearly 50 extra virgin oils, ranging in price from under $3 a bottle to over $20. I discovered that generally you get what you pay for — the more expensive the oil, the better the quality. Sadly, out of all the oils tasted, I found a great number of very bland or bad tasting oils. Listed below are some of my personal favorites.

SUPERMARKET BRANDS

Many people starting to use olive oil will probably pick up their first bottle in their local supermarket. There is no question that the extra virgins they sell are exceptionally cheap but I have found most of them to be disappointing, with flavors untypical of good olive oil. Many sell olive oil in plastic bottles, and in some cases I could detect a chemical backtaste which may or may not be from the bottle.

NATIONAL BRANDS

These are the olive oils produced by large companies and the ones you most commonly find on the shelves. Of the Italian brands, I think that Sasso extra virgin is the tastiest, it has a fresh olive smell with a green leaf taste. It is also exceedingly good value for money.

One of my most favorite olive oils and one I use a lot is Carbonell extra virgin from Spain. It is a richly fruity oil and is delicious for cooking and salads.

SINGLE ESTATE OLIVE OILS

These are the oils at the top end of the market in both quality and price. They are the "premier crus" and if you really want to discover what a fine olive oil should taste like you should treat yourself to a bottle of any of the following:

ITALY

Without question one of the best extra virgins from Italy is Fattoria Dell'Ugo, a light emerald green oil from Tuscany, with a lovely aroma of apples and a light peppery taste.

Equally I enjoy Olio Extra Vergine Di Oliva Della Riviera Ligure from Leonardo Raineri, a golden oil from Liguria with a light olive aroma and fruity, grassy flavor. Raineri also produce a Olio Extra Vergine Di Oliva, which is a pale gold with a light fresh taste and one which I think would be excellent for making mayonnaise, because it does not have an overpowering flavor.

From the Abruzzo region of Italy, an olive oil produced by Santagata is well balanced and fruity, golden in color.

Some other exceedingly fine Italian extra virgins to look out for are Badia a Coltibuono, Tenuto Carpazo, from Montalcino, Castello di Almonte from Umbria, and Colonna from the Molise region of central Italy.

FRANCE

Most of the French oils I tasted did not impress me, but it is only fair to say that there are few available in the stores and the best French oils are not available outside France. However Henri Bellon, who incidentally is the mayor of Fontvieille, produces a pale green, sweet fruity olive oil.

SPAIN

I personally like the style of Spanish olive oils very much. They are fruity and fragrant, without the peppery taste you get in some of the Italian oils. Two which are excellent are Lerida and Sierra de Segura from the Cooperativa Santa Ana.

GREECE

Some of the Greek style olive oils are wonderfully strong and fruity. My favorite is Sparta, made with organically grown olives.

WORLD PRODUCTION (1980/81)

		OIL (1000 TONS)			TABLE OLIVES (1000 TONS)		
COUNTRY	NO OF OLIVE TREES (MILLIONS)	PROD	EXPORT	IMPORT	PROD	EXPORT	IMPORT
Spain	193	416	92	–	151	78	–
Italy	183	407	17	95	74	2	22
Greece	113	211	20	–	63	40	–
Turkey	80	108	15	–	125	3	–
Tunisia	56	116	64	–	7	0.6	–
Portugal	50	42	2	2	2	20	–
Morocco	27	28	10	–	45	31	–
Syrian A.R.	20.4	47	–	–	32.6	–	–
Algeria	20	11.6	–	–	6.8	2.5	–
Argentina	5	30	4.8	60	64	20	–
France	4	1.7	6	26	2.4	2.5	27
USA	2.2	0.6	–	25	62	2	39.5

OIL AND FAT COMPOSITIONS

FOOD	% SATURATED	% MONO-SATURATED	% POLY-UNSATURATED
Coconut oil	92	6	2
Olive oil	12	80	8
Corn oil	16	27	57
Sunflower oil	10	18	72
Safflower oil	12	10	78
Butter	58	39	3
Margarine	64	30	6

INDEX
Recipes

Varieties of Olive

GENERAL INDEX

*Numbers in italic refer to
illustrations*